So Many Seeds

Learning the S Sound

Kerri O'Donnell

Rosen Classroom Books and Materials™
New York

There are so many seeds.

Some seeds are little.

Seed

Some seeds are big.

Some seeds grow in the dirt.

Seeds need some sun to grow.

Seeds need some water to grow.

13

Some seeds grow into flowers.

Some seeds grow into grass.

Some seeds grow into trees.

Some seeds grow into pumpkins, too!

Word List

seeds

so

some

sun

Instructional Guide

Note to Instructors:
One of the essential skills that enable a young child to read is the ability to associate letter-sound symbols and blend these sounds to form words. Phonics instruction can teach children a system that will help them decode unfamiliar words and, in turn, enhance their word-recognition skills. We offer a phonics-based series of books that are easy to read and understand. Each book pairs words and pictures that reinforce specific phonetic sounds in a logical sequence. Topics are based on curriculum goals appropriate for early readers in the areas of science, social studies, and health.

Letter/Sound: **s** – Prepare a poster with drawings or magazine pictures of items with initial consonant **s** (sailboat, salad, sandwich, seal, socks, sucker, soda, etc.). Pose riddles about the pictures. (Example: *I see something that begins with s. It lives in the sea.*) List responses on a chalkboard or dry-erase board.

- Give the child a response card with the letter **s**. Pronounce words with a variety of initial sounds. Have the child hold up their card only when they hear an initial **s** word. (Initial **s** words: *sad, safe, said, sat, saw, see, sap, seed, seven.*) List the words and have the child underline the initial **s** in each word as they read it. Ask them to use each word in a sentence.

Phonics Activities: Provide objects and pictures of **s** words (include a few that begin with other initial consonant sounds). Have the child name each picture or object and choose only those that begin with the initial **s** sound (*soap, sock, sandwich, salt, sun, seven, seeds, sand, sailboat,* etc.). Ask the child to place these in a box that is labeled *S Words* and is decorated with **s** pictures. Write the names of the **s** pictures and objects on a chart, underlining the **s** in each word.

- Write the following compound words on the chalkboard or dry-erase board: *sunbeam, sunset, suntan, sunflower, sunglasses, Sunday.* Have the child underline the part of each word that says *sun.* Pronounce the words in random order. Have the child use their knowledge of consonant sounds **b**, **s**, **t**, **f**, **g**, and **d** as an aid in identifying each word.
- Present additional compound words that have the root word *sun* (*sunlight, sunrise, sundress, sunburn, sunshine,* etc.). Discuss word meanings. Have the child illustrate one or more of the words for a chart entitled *Sunny Words.* Provide individual word cards to be used for an independent matching activity.

Additional Resources:
- Gibbons, Gail. *From Seed to Plant.* New York: Holiday House, Inc., 1991.
- Marzollo, Jean. *I'm a Seed.* New York: Scholastic, Inc., 1996.
- Medearis, Angela S. *Seeds Grow.* New York: Scholastic, Inc., 1999.

Published in 2002 by The Rosen Publishing Group, Inc.
29 East 21st Street, New York, NY 10010

Copyright © 2002 by The Rosen Publishing Group, Inc.

All rights reserved. No part of this book may be reproduced in any form without permission in writing from the publisher, except by a reviewer.

Book Design: Ron A. Churley

Photo Credits: Cover by Ron A. Churley; p. 3 © NomadicImagery/Moment Open/Getty Images; p. 5 © PRILL/Shutterstock.com; p. 7 © Dudarev Mikhail/Shutterstock.com; p. 9 © redmal/E+/Getty Images; p. 11 © Alf Ribeiro/Shutterstock.com; p. 13 © Elena Elisseeva/Shutterstock.com; p. 15 © bankrx/iStock/Thinkstock; p. 17 © Biletskiy_Evgeniy/iStock/Thinkstock; p. 19 © iStockphoto.com/backhanding p. 21 © LabyrinthZX/iStock/Thinkstock.

Library of Congress Cataloging-in-Publication Data

O'Donnell, Kerri, 1972-
  So many seeds : learning the S sound / Kerri O'Donnell.— 1st ed.
    p. cm. — (Power phonics/phonics for the real world)
  ISBN 0-8239-5908-2 (lib. bdg. : alk. paper)
  ISBN 0-8239-8253-X (pbk. : alk. paper)
  6-pack ISBN 0-8239-9221-7
  1. Seeds—Juvenile literature. [1. Seeds.] I. Title. II. Series.
  QK661 .O36 2001
  581.4'67—dc21
                                                2001000372

Manufactured in the United States of America